Bibliographic information published by the German National Library:

The German National Library lists this publication in the National Bibliography; detailed bibliographic data are available on the Internet at http://dnb.dnb.de .

Imprint:

Print and binding: Books on Demand GmbH, Norderstedt Germany
ISBN: 9783668899834

This book at GRIN:

https://www.grin.com/document/459457

Mila Zinkova

Air New Zealand Flight 901 and Polar Whiteout

GRIN Verlag

Flight into terrain, Air New Zealand Flight 901 and Polar Whiteout

Abstract

On 28 November 1979 on a sightseeing flight to Antarctica, the Air New Zealand Flight 901 crashed on the slopes of Mount Erebus. All 237 passengers and 20 crewmembers perished on the impact. Two investigations that followed both agreed that due to the Polar Whiteout the crewmembers of Flight 901 had not seen the mountain. This article offers a discussion on whiteouts in general and in regards to the accident. It explores other phenomena that could have led to the tragedy, namely spatial disorientation illusions.

1. Introduction

On the early afternoon on 28 November 1979, sightseeing Air New Zealand Flight 901 (TE-901) was approaching Ross Island, Antarctica. They were not going to land there. Instead, they planned to fly under a cloud base so that passengers could enjoy the wonders of the forbidden landscape below from the comfort of the plane. A low overcast was hanging over Ross Island. However, the Mac Centre advised TE-901 that below 2000 feet (around 600 m) the visibility is 40 miles (more than 60 km). Seventeen minutes before the crash, TE-901 captain Jim Collins requested permission to go below the cloud base in visual meteorological conditions (VMC), and the Mac Centre approved his request. Everybody thought that TE-901 was flying over flat sea ice of the McMurdo Sound. Nobody knew that actually TE-901 was approaching McMurdo station, flying over Lewis Bay, and that the plane was on a collision course with the active volcano Mount Erebus, the God of Darkness in Greek mythology.

The cockpit voice recorder (CVR) demonstrated that the TE-901 had descended from 17,000 feet (about 5,000 meters) to 1,500 feet (457 meters), but no one had ever seen a 12,444 feet (3794 meters) mountain directly in front of the plane. The investigators, the Chief Inspector of Air Accidents R. Chippindale (1979) and the Justice of the Royal Commission of Inquiry, Peter Mahon (1981), both attributed this invisibility to Polar Whiteout phenomena. However, Chippindale and Mahon understood the visibility-related consequences of Polar Whiteouts somewhat differently. Chippindale concluded that the accident was caused by descent in conditions of poor definition of the surface and horizon (as it usually is the case in the Polar Whiteouts conditions). Mahon (1981) argued that the Polar Whiteout conditions had deceived the pilots into believing that "the rising white terrain ahead was in fact quite flat and that it extended on for many miles". Mahon partly based his conclusion on the theory introduced by a pilot for Air New Zealand's Gordon Vette. Vette alleged that TE-901 encountered so-called "sector whiteout" (a Polar Whiteout that affects only one sector of terrain). Since then virtually every author writing about the tragedy (including a Weather Channel documentary "Sudden Impact" (Fraser, 2015)) adopted this "sector whiteout" theory.

This article offers scientific explanations and eyewitnesses' descriptions of Polar Whiteouts. It discusses Chippindale (1979) and Mahon (1981) reports. It provides some arguments that speak against the presence of a Polar Whiteout at the scene of the tragedy, and suggests other phenomena that could have caused the TE-901 pilots to be deceived into control flying into terrain.

2. Polar Whiteouts

The term "whiteout" is used to describe the lack of visibility arising in various meteorological conditions, including blizzards, fogs, etc. In this article, we will discuss the true whiteout (the Polar Whiteout) that is alleged to be responsible for the disaster.

The Polar Whiteout is, "a phenomenon of diffuse reflection and scattering by cloud and ground" Fritz (1957). The Polar Whiteouts occur

when clouds are so thick and uniform that light reflected by the snow is about the same intensity as that from the sky. No object casts a shadow, the surface of the snow merges into the sky, and no horizon, ground, or sky features can be recognized (Harker, 1958).

Writing in *Weather* Fritz (1957) emphasizes that true Polar Whiteouts are always associated with clear air:

It should be emphasized *that Polar Whiteout is independent of* any *ordinary meteorological obstructions to vision. It occurs in pure, clean air. Thus Stefansson (1945) states that ' under the conditions where a (polar) bear is invisible a hundred yards off, a blue fox might be visible at a mile or more '. The whiteout should not be confused with fog or haze; the air is clear during whiteout, but the combination of complete snow cover and overcast sky removes all chance of shadows or contrasts so that white objects are not discernible against their background, while dark objects can be seen readily.*

American military scientists have been very interested in studying whiteout conditions, and have tried to find ways to dispel whiteout-making clouds.

The U.S. Naval Institute (1956) writes:

Some observers believe that the light intensity is even "magnified" or increased during whiteout, *stating that the light reflects and re- reflects in all directions independent of the normal laws of reflection, refraction, and polarization,* which during white-*out seem to be inoperative.*

Schlichting et al. (1980) write:

When there is little in the environment except fields of snow and low cloud cover, the light is reflected back and forth until the distribution from all angles is uniform; in fact, physical measurements of the light, in a full 360 degrees around an individual reveal no differences whatsoever.

Some eyewitness accounts help us to understand how it feels to be caught in a Polar Whiteout. In her book "Alone in Antarctica", Aston (2014) writes:

There is no structure to focus on, no variation in shade or texture, no sense of distance, dimension or scale.

In his book “The Ice: A Journey to Antarctica” Pyne provides a rather poetic description of Polar Whiteouts:

[...] the whiteout—in metaphoric Russian, the white darkness. Loosely, a whiteout is any condition in which visibility vanishes. But more precisely, a whiteout occurs whenever light becomes so diffused, scattered, and reflected that all shadows disappear and the horizon as an identifiable feature is lost. Clouds at all levels can produce the effect, though high cirrus clouds are perhaps the most notorious because they appear so innocuous. The whiteout is the ultimate intensification of the reductionism that is the essence of The Ice. Light becomes an obscuring rather than an illuminating medium. Everything is brightened and nothing can be seen. All contrasts dissolve in a fog of light—a “milky, trembling nothingness”.

So hazardous is a whiteout that travel becomes impossible, but other properties make it the quintessence of Antarctic atmospherics. Unlike the blizzard, which is more common along the coast, whiteouts occur in the interior and are ominously passive, not active.”

In a true Polar Whiteout a person could see his own boots (if they are not white of course), but no surface he is standing at (Frost, 2011). In his book “Мир вертолета: Helicopter universe” Савинский (2014) quotes a Russian pilot who worked in Antarctica. He described the Polar Whiteouts conditions on the ground and in the sky:

A person in such conditions completely loses the ability to assess the scale of objects and the distance between them. A match stuck into the snow at a distance of only one or two meters from him seems to be a telegraph pole, moreover, it appears it is standing a kilometer from him.

[...]

Soon we reached the area [flying in a helicopter] *where the sky and snow merged into a single whole, and in the sky, as sometimes happen during the “whiteout”, several suns shone, and it was impossible to determine which one was real* [it appears the pilot was describing solar halos (Zinkova, 2018)].

Other authors describe the difficulty of determining the size of an object and distance to it under Polar Whiteout conditions:

Al described seeing an oil drum at some distance near White Island left by the British Commonwealth Antarctic Expedition. On close examination it revealed itself to be a five-inch ration box.

[...]

Someone else told of an article by Sir Raymond Priestley in which his colleagues had seen a driver and a dog team semaphoring at some distance, until the truth emerged. It was a tribe of penguins, *waddling away. And Captain Robert Falcon Scott is said to have once thought he saw a* dog team *that turned out to be a piece of film wiggling in the breeze* (Weihaupt et al., *2012).*

Writing in the *National Geographic* the famous polar explorer Richard E Byrd (1956) recalls how copter pilot Lt John Bacon described flying in the whiteout conditions:

In the air again, *the* "choppers" plunged into whiteout conditions, *the dread of polar flyers, when ground,* sky, *and horizon all are lost in a milky haze. "I felt like a fly trapped inside a ping-pong ball."*

It is rather rare to experience a scene with totally diffuse illumination. Usually some objects interrupt its milky harmony. The images presented in Figure 1 were all taken in Antarctica, and although none shows either horizon or shadows and there are hardly any differences between the colours of the sky and the ground, none of them depicts a true Polar Whiteout.

Figure 1. Upper frame: Tourists are exploring an iceberg that was frozen in the sea ice in the Ross Sea, middle frame: Colony of Adélie penguin at Cape Royds; lower frame the memorial Cross at Wind Vane Hill (Cape Evans) with Robert Falcon Scott's Discovery Hut. Both images are the author's own work.

The image presented in Figure 2 is not an image of a Polar Whiteout either. This is a picture of a lighthouse surrounded by fog, photographed in San Francisco. However, this image could serve as a good visualisation of the signatures of Polar Whiteouts we discussed above (except that this *white* lighthouse would not be visible under Polar Whiteout conditions), namely the indistinguishable horizon, the same colour of the sky and the ground, and the absence of shadows. In addition, it is impossible to determine how far or how large the structures are. In the absence of any other visual landmarks, it is impossible to say whether these structures are toys located a few meters from the camera, or real buildings located a few kilometres away.

Figure 2. Point Benito Light House in fog. The author's own work.

3. Meteorological condition in the area of the disaster

In his report Chippendale (1979) relisted the meteorological conditions in the area of the disaster. General situation was described as follows:

O*n 28 November 1979 at 0100 hours (Z)* [11 minutes after the accident] *the McMurdo area was under the influence of a surface low pressure* [...] *the surface definition at the time* [was reported] *as poor and the horizon definition fair. Mountain tops in the area were covered in cloud.*

None of the flight crews located in the immediate vicinity of the disaster zone described the conditions as a Polar Whiteout. Eleven minutes after the crash an American Navy plane was located around 80 miles (around 128 km) from McMurdo. They reported:

[...] *continuous stratoform layer covering Ross Island with cloud "domes" over Mt Erebus and Mt Terror which concealed the mountains from view. The cloud layer extended to the north of Ross Island. A lenticular "cap cloud" was over Mt Erebus above the main cloud layer."* [...]

The lowest layer was solid overcast with a ragged base. The visibility was good below the cloud base but the surface definition was poor (Chippendale ,1979).

One hour eleven minutes after the crash a helicopter

landed on the beach 10 km from the accident site and the conditions at that time were overcast with light snow but the sun could be "made out" through the cloud occasionally. The surface definition at the time was very poor to nil. [...]*They were not able to distinguish that the slopes to the south* [the slopes of Erebus] *were elevated or separately identify cloud and snow. No bare rock was visible on the slopes but the rocky coastline below the ice cliffs was visible* (Chippindale ,1979).

A United States Air Force was following approximately 45 minutes after TE-901. The captain of this aircraft described the clouds as "ordinary cumulus or stratocumulus" with a good visibility below the base. (Chippendale ,1979).

Opinions vary widely on what kind of clouds give rise to Polar Whiteouts. Fritz (1957) appears to agree with Hedine that Polar Whiteouts occur with cirrostratus and/or altostratus clouds. Other authors state that low-level stratus clouds could make whiteouts as well. Most authors agree that Polar Whiteouts can occur under any level of stratus overcast. Cumulus and/or stratocumulus as described by an American pilot do not look promising for the formation of a Polar Whiteout. Furthermore, the helicopter's pilot could not see bare, black rocks on the slopes of Erebus, which probably means they were behind meteorological clouds. At the same time, lenticular clouds (Figure 3) are common in Antarctica, especially around Erebus and they cast deep shadows https://antarcticfudgesicles.files.wordpress.com/2010/11/dsc_1054.jpg. The presence of lenticular cloud might have prevented a Polar Whiteout from forming.

Figure 3. Lenticular clouds in Antarctica. The author's own work.

4. CVR

In this section, we will discuss visibility-related statements made by the crewmembers in the cockpit in the last minutes of the flight.

All times below are in UTC.

The plane crashed at 0049:50.

0047:06 captain: "Down to two thousand feet."

At about the same time, or later, a passenger took a picture http://photobucket.com/gallery/user/turricaned/media/cGF0aDovaWVfcGFzc2VuZ2VyX3NiX2Rlc2MucG5n/?ref= Chippindale (1979) described the weather associated with the image:

Over Northern Victoria Land the weather was clear with an almost complete absence of cloud at any altitude. The aircraft flew over continuous cloud layers from about Franklin Island to just north of Beaufort Island where the aircraft was able to descend through an obvious break in the cloud cover. Several photographs show a clearly defined cloud base beyond and above the Lewis Bay coastline of something less than 2000 feet.

0047:43 Captain: "We might have to pop down to fifteen hundred here I think". Second pilot: "Yes, OK. Probably see further in anyway".

0047:49 Second pilot: "It's not too bad".

0048:46 Captain: “Actually those conditions don’t look very good at all – do they?” Somebody: “No, they don’t.”

0049:08 Tour guide: “That looks like the edge of Ross Island there”.

0049:30 Captain: “We’re twenty six miles north we’ll have to climb out of this”.

0049:33 Second pilot: “It’s clear on the right and (well) ahead”. Captain “is it?” Second pilot: “yes”.

0049:35 Tour guide: “You can see (Ross Island) Right Fine!”

0049:38 Second pilot: “Yes you’re clear to turn right there’s no high”. Captain: “No, negative.” Second pilot: “No high ground if you do a one eighty”.

In his report, Chippindale (1979) described the last seconds of the flight before the impact:

After the captain’s decision to climb the aircraft out of the area he and the co-pilot were discussing the most suitable climb out path when the ground proximity warning system sounded instructing the crew to “Pull up”. The crew responded to the alarm without undue hesitation, the flight engineer calling off the heights of 500 and 400 feet indicated on the radio altimeter and the captain calling for “Go-round power”. The warning 6½ seconds before the impact was, however, too late for the crew’s action to make any significant effect on the aircraft’s level flight path. Their reaction time was established as very similar to or better than that of experienced crews placed in a similar situation in the training environment of the flight simulator. It is likely however that as a result of a whiteout the go-round attempt was procedural in response to the warning rather than a desperate attempt to avoid a readily apparent obstacle.

A pilot for Air New Zealand Arthur Cooper listened to the original record and he stated that he has heard no alarm in any voice up to the end (Cleal, 2009).

The weather conditions in Antarctica can turn from a clear horizon definition to a no horizon definition in a matter of seconds. In 2015, an Australian helicopter crashed in the ice shelf, when suddenly the horizon definition was lost and surface definition was reduced. However, the helicopter crashed after the pilot sounded an alarm, and tried to turn back (Commonwealth of Australia, 2015). The absence of alarm in the TE-901 situation indicates that the crew did not experience the loss of visible horizon at any time.

5. Chippendale's report

A map of Ross Island and adjacent areas with an approximate path of the actual TE-901 trajectory (Figure 4) shows that TE-901 was approaching Ross Island, flying over Lewis Bay directly in the direction of Erebus. Pilots thought they were flying west of Ross Island over flat sea ice in the McMurdo Strait (Figure 5).

Figure 4. The map of Ross Island by USGS (http://www.mappery.com/map-of/Ross-Island-Topo-Map accessed on 5 January 2018) modified by the author to show the path of the flight. Approximate path of

TE-901 towards Erebus is shown in red colours. The image is in public domain because it is the work of US Government

Figure 5. McMurdo Sound with two icebreakers and a few seals. The author's own work.

In his report, Chippendale (1979) wrote:

The weather conditions as described had a high potential for a "whiteout" the phenomenon which is always likely when overcast conditions exist above a continuous snow covered slope

[...]

Micro-summary: This McDonnell-Douglas DC-10 collided with a mountain during a sightseeing charter in whiteout conditions.

However, as we discussed above, not every overcast sky produces a Polar Whiteout.

Chippindale (1979) included in his report a detailed diagram of the flight path (Figure 6) with the distances and heights of the terrain and the plane.

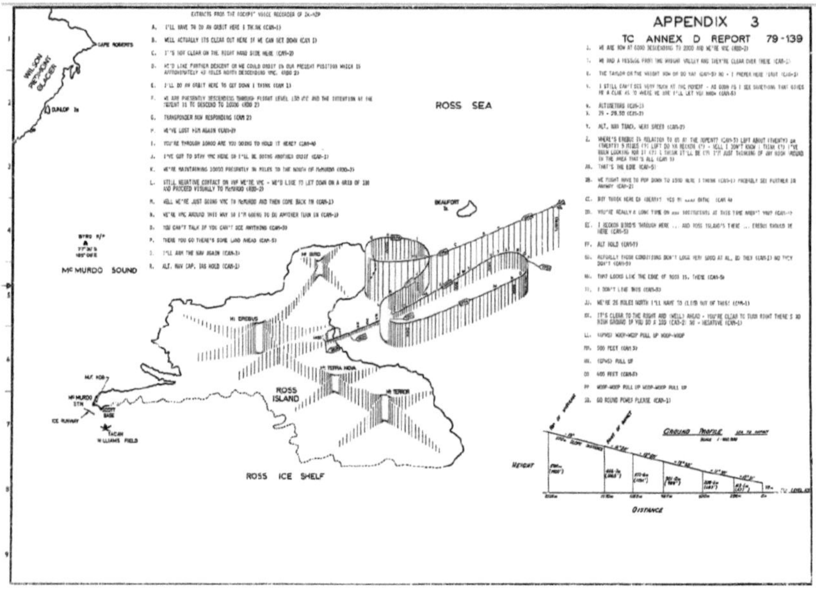

Figure 6. The flight path of TE-901 with Beaufort Island located on the right- hand side of the path by Transport Accident Investigation Commission. The image is in public domain because it is New Zealand's Government work. http://www.erebus.co.nz/portals/4/documents/reports/chippindale/79-139_section1.pdf (Accessed on 11 March 2019).

The diagram shows TE-901 made two descending turns, which means that the passengers and pilots were able to see in every direction, including the direction of an alleged sector whiteout. However, it appears that none of the photographs they took depicted a Polar Whiteout in any sector of the sky:

In a number of other cases the last photographs (some taken only seconds before the collision) show the eastern shoreline of Cape Bird and the north eastern and north western coastline of Lewis Bay [see Figure 4] *and a cloud layer with a base of some 2000 feet, above an unbroken snow covered slope* (Chippindale, 1979).

An important landmark Beaufort Island was clearly distinguishable on some photographs:

The aircraft flew for some time within sight of Beaufort Island, which was clearly visible. The sun was shining on the north eastern slopes of Mt Bird with rock outcrops and the ice cliff face around this section of Lewis Bay clearly visible (Chippindale, 1979).

If the pilots realized that the island should have been on the left-hand side, they would have understood they were not on the right path. Vette alleged that the crew mistook the Island with a patch of open water. If they did, this may be just another demonstration of how easy it is not to see something so obvious even when the visibility is perfectly clear.

Chippindale (1979) concluded:

The probable cause of this accident was the decision of the captain to continue the flight at low level toward an area of poor surface and horizon definition when the crew was not certain of their position and the subsequent inability to detect the rising terrain which intercepted the aircraft's flight path.

6. Captain Gordon Vette's research

Vette and many others disagreed with Chippendale's (1979) conclusion. To justify Collins's decision to descend and fly low in VMC Vette had to demonstrate that the visibility *appeared* to be good and Collins was sure he was flying over McMurdo Sound. This optical illusion was made possible by a "sector whiteout", Vette alleged. Writing in Smithsonian Air & Space Negroni (2015), credits Vette as "the man who solved the mystery of Flight 901":

Vette's alternative explanation was considered credible, and was incorporated into the commission's final report, which was officially accepted by the government in 1999. Vette's theory of sector whiteout, where visibility is affected in just one direction, is now an accepted factor in arctic flying.

During his research, Vette had a few pieces of evidence at his disposal, namely CVR, passengers' pictures, and an artist representation of what Collins and his crew would have seen in sector whiteout conditions (http://www.erebus.co.nz/Portals/4/Documents/Reports/Mahon/Whiteout%20Phenomenon.pdf scroll down to see the sketches). Although a few pictures taken in the last moments were "considerably damaged by the impact forces", in his book "Impact Erebus", Vette (1984) described the one that was taken just 51 seconds before the impact:

Shadows on the sea ice show *the cloud above was of variable density. There are fewer bright patches on the surface of the right of the photo indicating that there was a denser cloud in this area, towards the impact area. This would be the direction that Collins and Mulgrew* [the guide] *would be examining closely for signs of buildings, etc in the McMurdo-Scott Base area as the edge of what they believed was Cape Royds moved back behind them to give them an unobstructed view of this part of Ross Island.*

The images in Figure 7 show some areas Vette mentioned in the above description.

Figure 7. Upper Frame: Cape Royds with Sir Ernest Shackleton's Hut in the foreground. Middle frame: McMurdo Station. Lower Frame Scott Base. All images are The author's own work.

According to Vette's description of the cloud in the passenger's picture, the cloud was of "a variable density" and there were some shadows on sea ice. Not uniform overcast results in the combination of bright and shadowy areas (Figure 8). Denser clouds over some areas do not prevent the low Antarctic sun from lighting up the terrain below (see Figure 8 the lowest frame). Vette's description of the passenger's picture does not appear to be consistent with the description of a Polar Whiteout.

Figure 8. Not uniform overcast results in the areas with a different amount of illumination. All images were taken in Antarctica. All images are the author's own work.

In his book, Vette (1984) described the meteorological conditions that he believed led to the tragedy:

Unknown to themselves, Collins and his crew were confronted with cloud concealing Mt Erebus at the top of their view and fog concealing ice cliffs at the bottom of their view while the sun behind them produced no shadows discernible by them. Multiple light reflection back and forth between snow and cloud created whiteout, rendering Erebus invisible and probably resulting in a false horizon [...]

Vette's statement is rather confusing: if the terrain was covered by cloud above and fog below, both are perfectly capable of hiding a landscape on their own on a clear day (Figure 9), then what role the whiteout played in "rendering Erebus invisible" and by what means whiteout resulted in a false horizon? In addition, "fog at the bottom of their view" (ground fog) would have greatly reduced the reflectivity of snow, and Fritz (1957) wrote that "*whiteout depends entirely on the presence of a very highly reflective snow surface*".

Figure 9. Lenticular cloud and fog over Golden Gate Strait. Imagine this image with everything covered with snow. The author's own work.

7. Mahon's "whiteout"

After Chippendale's report that cited pilot error created a controversy, another investigation was ordered. Justice Mahon was selected to conduct this investigation.

A year after the tragedy Mahon visited the crash site. While flying in a helicopter towards the site Mahon and his team observed how Erebus disappeared behind clouds. The pictures they took looked surprisingly similar to Vette's artist representation that we discussed earlier, including the presence of a false horizon. Vette declared that Mahon observed a sector whiteout. However, the description of the meteorological conditions that Justice Mahon (1985) encountered during the flight leaves little doubt that no Polar Whiteout was present:

The pale cloud drifting from Mount Bird towards Mount Erebus unexpectedly increased in height and within a minute or so the massive structure of Mount Erebus totally disappeared. This was certainly a striking illustration of the unpredictable weather patterns of Antarctica.

Then we noticed an ice fog [forming of ice fog is very unlikely, probably Mahon observed just regular ground fog] *lifting off the solid ice in front of us. It was drifting over the ice shelf* [he probably meant a glacier on the slopes of Erebus] *and entirely concealing three or four miles of it to our left.*

Mahon's whiteout has the same issues as Vette's whiteout. The combination of drifting meteorological cloud that concealed Mount Erebus and ground fog is inconsistent with the preconditions that are necessary for a Polar Whiteout to form. In fact, clouds and/or ground fog are indistinguishable in the presence of any whiteout at all. In addition, Polar Whiteouts are clear air whiteouts that have nothing to do with fog.

In his report, Mahon (1981) acknowledged that Chippindale did indeed conclude that the whiteout conditions were present. However, Mahon did not agree with Chippindale that whiteout created "poor surface and horizon definition".

Mahon devoted 14 pages of his report into the accident to the whiteout phenomenon http://www.erebus.co.nz/Portals/4/Documents/Reports/Mahon/Whiteout%20Phenomenon.pdf. He interviewed numerous experts. Vette also testified. One of Mahon's experts, Mr. Shannon, vice-president of operations for Bradley air services Limited of Ontario, said that he regarded the possibility of false horizon as being remote, and yet he "had no doubt that a pilot unfamiliar with Polar conditions would believe that he was flying forward with clear visibility over flat terrain for many miles" (Mahon 1981). In his vision of Polar Whiteouts Mahon (1981) sided with Shannon and some experts:

It became apparent that *although* the *aircraft was flying* in *clear air*, *not one of the five persons on the flight deck ever saw the mountain side with which the aircraft collided.*

It was quite apparent that the air crew had been deceived into believing that the rising white terrain ahead was in fact quite flat and that it extended on for many miles.

However, neither Mahon nor his expert explained how a Polar Whiteout could have deceived somebody into believing that the rising terrain ahead was flat and extended on for many miles. Sometimes, if some familiar dark objects are visible, a person might try to estimate distances in Polar Whiteout conditions. However, if according to Mahon there were no any visible objects ahead of TE-901, they would not have been able to estimate the distance. For example, who could tell how many miles of clear visibility over flat terrain there is in the lowest frame of Figure 7 (and it is not even a true Polar Whiteout)? Here is the same terrain in a *good* visibility https://thumbs.dreamstime.com/z/scott-base-ross-island-antarctica-sign-main-new-zealand-pressure-ridges-sea-ice-can-be-see-foreground-73442080.jpg . It is difficult to imagine that a pilot (even unfamiliar with Polar Whiteout conditions) could be deceived into believing that he has many miles of clear visibility over flat terrain if the sky and the ground are merged in whole, no horizon is visible and ahead there is only "milky, trembling nothingness". In fact, Harker (1958) describes what happens, when one tries to see in an empty visual field:

Continued efforts "to see" in such an empty visual field usually result in excessive tearing, a sense of eye strain and, very probably, dissociation of the eyes.

There are no signs that the crewmembers have experienced any of the above, never once they sounded the alarm, which means that they really believed they were flying over even terrain that extended for many miles ahead. As was demonstrated above a Polar Whiteout could not have been the cause of such a delusion, meaning that the TE-901 encountered other phenomena.

8. A false horizon

Many dangers are associated with flying over unfamiliar, mountainous terrain even in sunshine. In fact, one of the Mahon's (1981) experts said:

In snow-covered terrain, a pilot is deprived of texture information which will alone acquaint him with slope and distance. In bright sunshine he is only deprived of that information to a partial extent. But even so, his normal appreciation of variations in terrain is adversely affected.

In addition, to usual problems with low flying over snow-covered, unfamiliar mountains the crew of TE-901 probably experienced a false horizon. A false horizon seems to be a permanent feature in the area of the disaster and is due to the upsloping terrain itself (Figure 9). Although on the day of the tragedy a meteorological cloud obscured the top of Erebus the pilots still could see that false horizon.

Figure 9. "Looking southwest past burnt out cabin towards Mt. Erebus." Source Chippindale's (1979) report http://www.erebus.co.nz/portals/4/documents/reports/chippindale/79-139_section1.pdf (Accessed on 11 March 2019). The image is in public domain because it is New Zealand's Government work.

If Mahon appeared to agree with his expert in regards to the undetectable horizon, Vette has always maintained that a false horizon was visible. As a pilot himself, he understood that, if there was poor, or nil horizon definition, Collins, as a professional pilot, should not have flown the plane in VMC at 500 metres high in very close proximity to the Island that hosts some high mountains. In fact, Vette was credited as a person who first described sector whiteout as "a modified whiteout in which the viewer can distinguish the horizon line but still cannot see the obstruction in front of him" (Greene, 1988). However, there are no scientific studies or eyewitness' accounts of such an event. A false horizon and a true Polar Whiteout are two mutually exclusive phenomena.

A false horizon needs no whiteout to deceive a pilot. A false horizon could cause an observer to overestimate the distance to it. For example, how far is the horizon from the camera in the image presented in Figure 10?

Figure 10. False horizon. The author's own work.

In fact, the image presented in Figure 10 is a crop of the image presented in Figure 11. The false horizon is around 1 km away from the camera.

Figure 11. Fog in Golden Gate Strait. The author's own work.

In addition, a false horizon causes upsloping terrains to appear horizontal due to a common spatial disorientation illusion:

Sometimes the horizon perceived through ambient vision is not really horizontal. Quite naturally, this misperception of the horizontal creates hazards to flight. A sloping cloud deck, for example, is very difficult to perceive as anything but horizontal if it extends for any great distance into the pilot's peripheral visual field. Uniformly sloping terrain, particularly upsloping terrain, can also create an illusion of horizontality with disastrous consequences for the pilot thus deceived (Gillingham and Previc, *1993*).

A false horizon illusion could be deadly:

This false horizon illusion can be dangerous if the aircraft is operating at speed and low altitude. If unrecognised, the situation can lead to a ground impact relatively quickly.

Featureless snow-covered terrain could be responsible for another spatial orientation illusion, namely black hole illusion (BHI). BHI creates the appearance of being higher above the ground than one really is:

The BHI can be thought of as a specific form of a featureless terrain illusion. The United States Department of Transportation (US DoT) defines the featureless terrain illusion as: An absence of ground features, as when landing over water, darkened areas, and terrain made featureless by snow, can create the illusion that the aircraft is at a higher altitude than it actually is (Gibb, 2007).

9. Conclusion

The meteorological situation at the time of the disaster with clouds of variable density, the breaks in the clouds, the sun lighting up some areas and shadows on the ground excludes the presence of a Polar Whiteout. The summit of Mount Erebus was concealed by meteorological clouds. The visibility was not very good, but a false horizon tricked the pilots into believing the visibility was fair. It appears that the crew of TE-901 encountered the conditions similar to those Mahon observed during his visit to the crash site, and as we discussed earlier, no Polar Whiteout was present during his flight.

The article has offered a new solution that explains what could have tricked the pilots into control flying into terrain, namely the combination of clouds concealing the summit, featureless snow-covered terrain and a false horizon that created some deadly visual illusions associated with spatial disorientation.

I was not able to find neither any scientific research on sector Polar Whiteouts nor any eyewitnesses' accounts about observing one. Some meteorological whiteouts could occur only in one sector of the horizon. However, true Polar Whiteouts with their diffused reflections between uniformly overcast sky and uniformly snow-covered ground might be undetectable to an observer located outside. Maybe to experience a true Polar Whiteout an observer should be fully engulfed in it. Even if a Polar Whiteout could be observed in only one sector of the sky, the horizon definition in this sector would be nil.

10. Acknowledgements

The author thanks polar explorers and scientists Gary Lewis, Dr. David Harrowfield and Robert Mckerrow for helpful email discussions in regards to the Polar Whiteouts and Flight 901.

References

Aston F. 2014. Alone in Antarctica: The First Woman To Ski Solo Across The Southern Ice. *Counterpoint.* Berkeley, California.

Byrd RE. 1956. *The National Geographic Magazine.* **110**: 153.

Chippendale R. 1979. Flight into terrain, Air New Zealand McDonnell-Douglas DC-10-30 ZK-NZP, Ross Island, Antarctica, November 28, 1979. http://www.erebus.co.nz/portals/4/documents/reports/chippindale/79-139_section1.pdf (Accessed on 11 March 2019).

Cleal W. 2009. Flight transcript: The moment of impact. http://www.stuff.co.nz/national/erebus/3030756/Flight-transcript-The-moment-of-impact. Accessed on 7 October 2018.

Commonwealth of Australia. 2015. Controlled flight into terrain involving Aérospatiale AS350B2 VH-HRQ. *Australian Transport Safety Bureau.* *http://www.atsb.gov.au/media/5567885/ao2013216-final_report.pdf*. Accessed on 7 October 2018.

Gillingham KK. and Previc FH. 1993. Spatial orientation in flight. Armstrong Laboratory (AFMC). *Brooks Air Force Base.* Texas. USA. http://www.dtic.mil/dtic/tr/fulltext/u2/a279306.pdf. Accessed on 7 October 2018.

Gibb RW. 2007. Running Head: BLACK HOLE ILLUSION. http://www.dtic.mil/dtic/tr/fulltext/u2/a462899.pdf Accessed on 7 October 2018.

Greene HH. Beattie v. United States, 690 F. Supp. 1068 (D.D.C. 1988). https://law.justia.com/cases/federal/district-courts/FSupp/690/1068/2359221/ Accessed on 7 October 2018.

Fritz S. 1957. THE 'POLAR WHITEOUT'. *Weather.* **12**: 345-349.

Frost S. 2011. Whiteout: Schneefälle und Weißeinbrüche in der Literatur ab 1800 (Lettre). *transcript Verlag.* Bielefeld, Germany.

Hansen WR. 1994. Greenland's Icy Fury (Williams-Ford Texas A&M University Military History Series). *Texas A&M University Press.* College Station, Texas, USA.

Harker GS. 1958. Whiteout—A Bibliographical Survey. *Bulletin American Meteorological Society.* **40**:225-228.

Mahon PT. 1981. Royal commission to inquire into and report upon the crash on Mount Erebus, Antarctica, of a DC10 aircraft operated by Air New Zealand Limited. http://www.erebus.co.nz/Portals/4/Documents/Reports/Mahon/Mahon%20Report_web.pdf . Accessed on 7 October 2018.

Mahon P. 1985. Verdict on Erebus. *Collins.* Auckland, New Zealand.

Negroni C. 2015. The Man Who Solved the Mystery of Flight 901. https://www.airspacemag.com/daily-planet/man-who-solved-mystery-flight-901-180956222/ Accessed on 7 October 2018.

Newman DG. 2007. An overview of spatial disorientation as a factor in aviation accidents and incidents. *Australian Transport Safety Bureau.* Canberra City, Australian Capital Territory.

Pyne SJ. 2017. The Ice: A Journey to Antarctica (Weyerhaeuser Environmental Books) Kindle Edition. *University of Washington Press.* Seattle, WA, USA.

Rogers J. 2018. Torrential rain from Hurricane Lane caused steamy white-out conditions at Hawaii's Kilauea volcano. *FoxNews. https://www.foxnews.com/science/hurricane-lane-created-a-steamy-white-out-at-hawaiis-kilauea-volcano.* Accessed on 7 October 2018.

Савинский ЮЭ. 2014. Мир вертолета.: Helicopter universe. *Триум.* Moscow, Russia.

Schlichting CL, Luria SM., Kinney JAS, et al. 1980. Aids for improving vision in white-out. http://www.dtic.mil/dtic/tr/fulltext/u2/a088726.pdf Accessed on 7 October 2018.

Vette G. 1984. Impact Erebus. *Sheridan House Inc.* Dobbs Ferry NY United States

Weihaupt JG., Stuart AW., Van der Hoeven FG., et al. 2012. Impossible Journey: The Story of the Victoria Land Traverse 1959-1960, Antarctica. *Geological Society of America.* Washington, DC. USA.

U.S. Naval Institute. 1956. United States Naval Institute Proceedings **82:** 957.

Zinkova M. 2018. Ice halos in weather lore and meteorology, and a rare halo display in San Francisco. *Weather.* **73:** 239-246. doi:10.1002/wea.3237